Geology in the Elementary School

The active enthusiasm of children provides a great opportunity to start building basic geological concepts and vocabulary about landforms. What better place than a sandbox to learn about such things as:

Landforms

mountain
hill
mountain range
foothill
plateau
volcano
valley
pass

Waterways

river
pool
lake
swamp
stream
brook
rapids
waterfall

Coastlines

estuary
delta
ocean
bay
peninsula
island
archipelago

Display pictures of each of the various types of landforms and waterforms around the classroom.

Materials You'll Need

Your regular playground sandbox will work fine, preferably with wet sand from a recent rain, hose, or bucket.If your playground doesn't have a sandbox, you can make a temporary one by filling a plastic wading pool with sand. Some teachers have a sand table right in their classroom. You'll also need:

- source of water: a hose or container
- beach buckets and shovels
- milk straws to hold up landform signs
- assorted rocks and gravel
- carpet remnants or doormats
- 1' - 2' pieces of wood or cardboard
- pieces of plastic to put under "lakes"
- ice cubes

Opening Activity

Begin by finding out what your students may or may not already know. Head for the sandbox. Select some students at random. Ask them to make examples of the geological features you name in the sand. Assure them that making a mistake will not be a bad thing to do. Your purpose is to find out what you need to teach. Ask the selected students to make these features and see what you get:

- mountain
- plateau
- mountain range
- mountain pass
- coastline with a bay, a peninsula, and an archipelago

For the teacher

Making Landforms

Let's start out with the basic "ups and downs" of the land.

1. Discuss these eight basic landforms:

• **mountain**	a mound of land 1000 ft or more above the surrounding land
• **hill**	a mound of land that is not as big as a mountain - less than 1000 ft high
• **mountain range**	mountains in a row and connected so that they don't appear to be separate mountains
• **foothills**	a row of hills along the base of a mountain range but not as tall
• **plateau**	a mountain or hill that is flat and level on top
• **volcano**	a mountain with a hole in the middle where the lava came out to form the mountain
• **valley**	the low area between two mountain ranges

2. Label a tagboard card for each landform. Cover it with contact paper and tape it to a straw.

3. Demonstrate for your students how they can make a landform in the sand: (Use 6" (30 cm) to represent 1000 feet. A hill will be less than 6" high.)
   - Read the definition for "hill."
   - Make a hill in the sand.
   - Stick the "hill" sign into the hill you made.

4. Have students demonstrate the other seven basic landforms.
   - Select a student for each remaining landform.
   - Read the definition for each landform, one at a time.
   - Have the assigned student make it in the sand.
   - Have the student put on the appropriate sign.
   - Ask each student to repeat the name of what he/she made and its definition.

5. What will the other students do while the landforms are being created?
Have the students who were not selected for this first group make notes, using the reproducible student sheet on page 4. Print the sheet on both sides. Students will need a book , magazine, or a cardboard "clipboard" to put their paper on while writing. Keep these observation notes throughout the unit. They can form the basis for a geology log or portfolio.

The students:
- fill in the name of each landform
- fill in the description/definition
- make a sketch of the landform

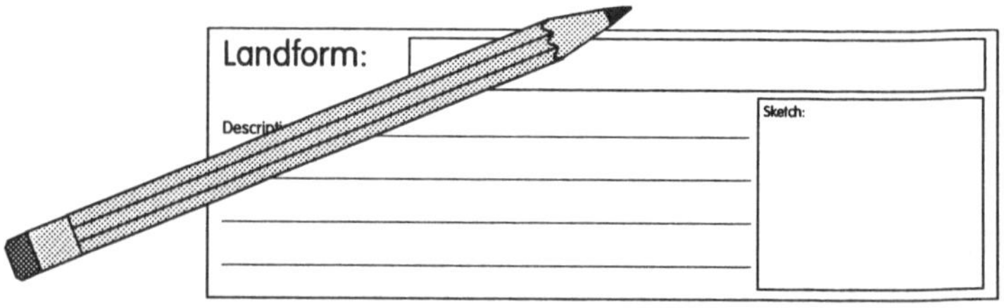

Let them know that everyone will have more than one turn in the demonstration group. Ask students in the observation group, "What is this landform? How do you know? Why isn't that one called the same thing?" etc.

Card Games

There are two uses for the reproducible cards on page 3, 11, and 13.

1. Cut the cards apart and use them to fill in the blank labels on the colored poster.
2. Cut apart and scramble the cards and arrange them into matching sets of three.

Note: these cards along with poster activities or as an independent matching activity (see page 2).

mountain	a mound of land 1000 ft or more above the surrounding land	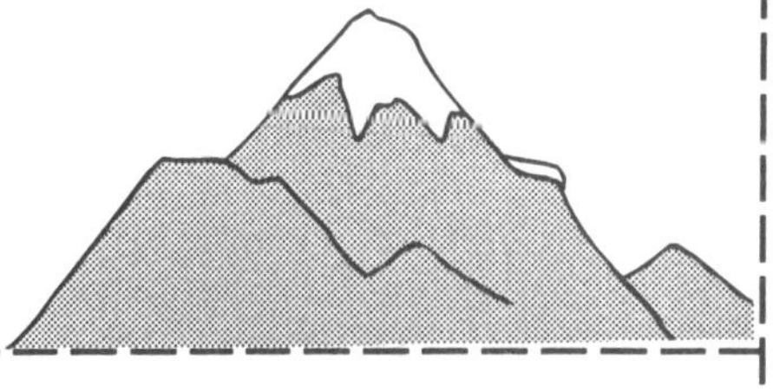
hill	a mound of land that is not as big as a mountain -less than 1000 ft high	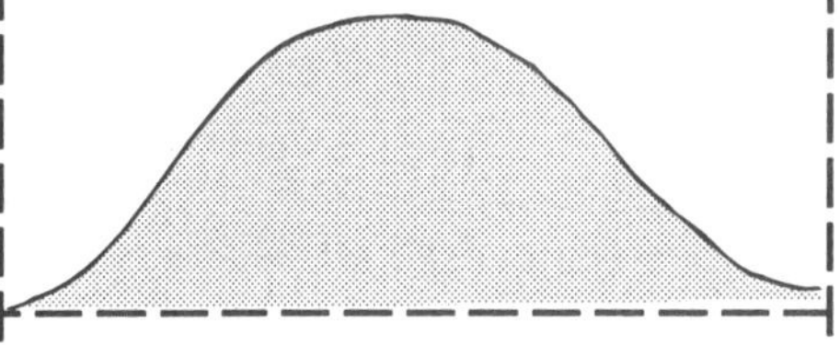
mountain range	mountains in a row and connected so that they don't appear to be separate mountains	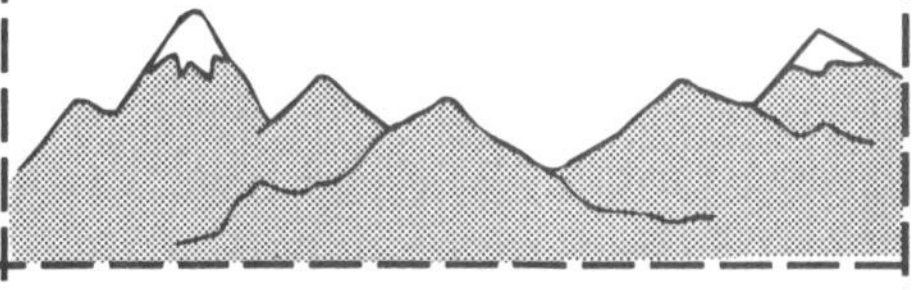
foothills	a row of hills along the base of a mountain range but not as tall	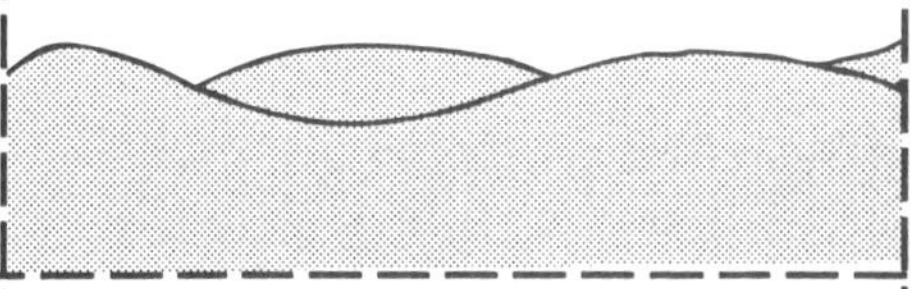
plateau	a mountain or hill that is flat and level on top	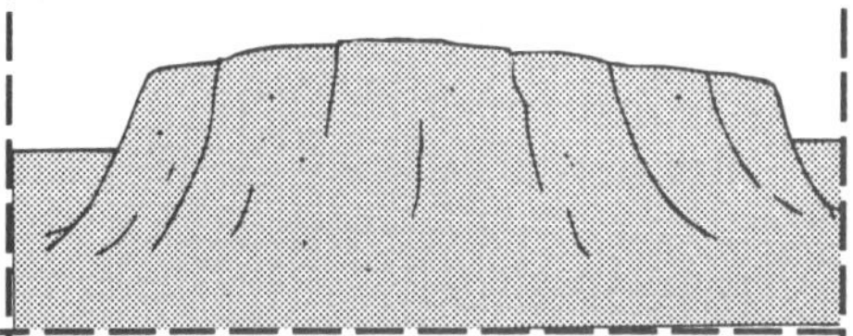
volcano	a mountain with a hole in the middle where the lava came out to form the mountain	
valley	the low area between two mountain ranges	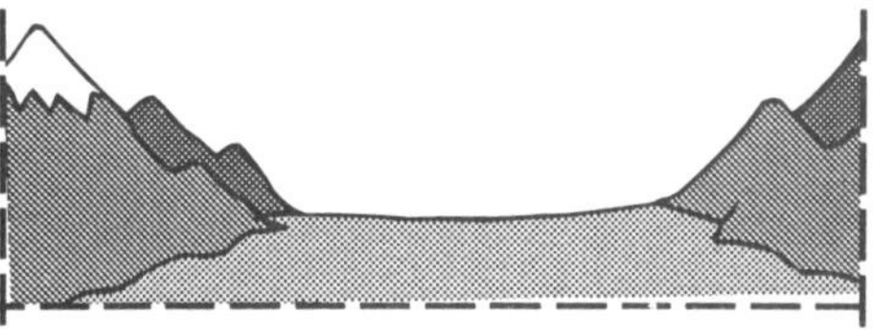
pass	a low spot between two mountains in a range where it is easier for people to get through	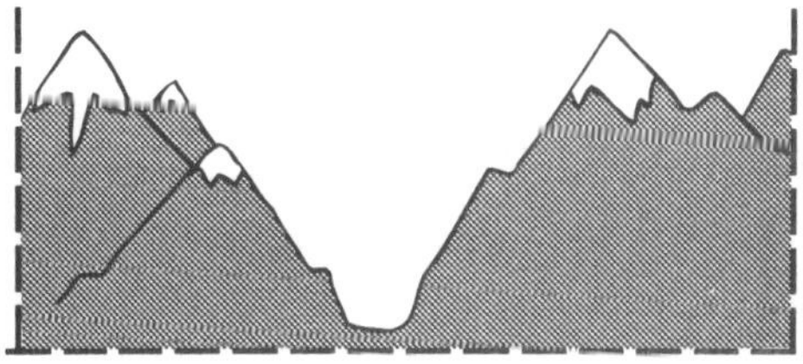

Landform:

Description:

Sketch:

Landform:

Description:

Sketch:

Landform:

Description:

Sketch:

Landform:

Description:

Sketch:

How Did Those Mountains Get There?

Mountains are formed in three ways in nature.

Select a group of students to demonstrate in the sandbox as you narrate for them and the observers.

1. *Folded* mountains are smooth.

As the earth cools, its "skin" gets to be a little too big for the shrinking earth, so the earth's crust folds, producing mountain ranges.

Put an old piece of remnant carpet or a doormat just under the surface of the sand. Then, have two students push the opposite ends of the carpet toward each other, causing the carpet or mat to fold upward in the middle, simulating the folded mountain range. If the sand falls off the carpet, have them recover it.

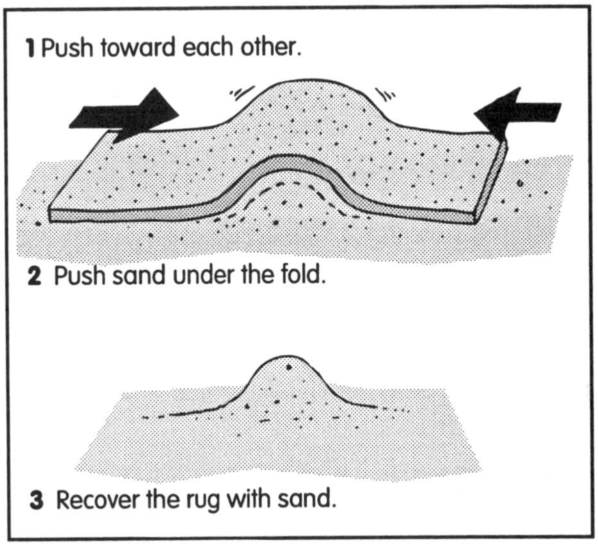

2. *Block* mountains are jagged.

(until weathering anderosion take over)

As the earth cools, cracks sometimes appear in the crust. They are called faults. Pressures that would cause a fold produce a different result where there are faults. Instead of folding, the two sides of the crack move separately. When one side goes up more than the other, a sharp ridge is formed.

Put two pieces of straight-edged wood under the sand. Wiggle both boards (causing an earthquake). Tilt the board on one side higher than the one on the other side, simulating a block mountain range.

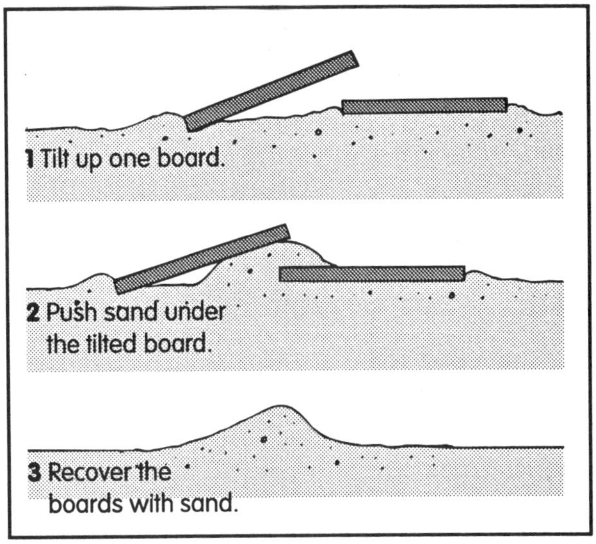

3. *Volcanic* mountains are cone-shaped.

Even though the earth's surface is cooling, there is still lots of molten (melted) hot rock under the surface called magma. When pressure builds up, the magma can be pushed out onto the surface. When it comes out it is called lava. The lava collects on the surface around the hole it came up through and forms a cone-shaped mountain called a volcano. Volcanoes tend to be scattered randomly, so they don't form straight ridges or ranges like folded and block mountains do.

Make cone-shaped mountains with holes in the middle where the lava comes out.

How about some sound effects?

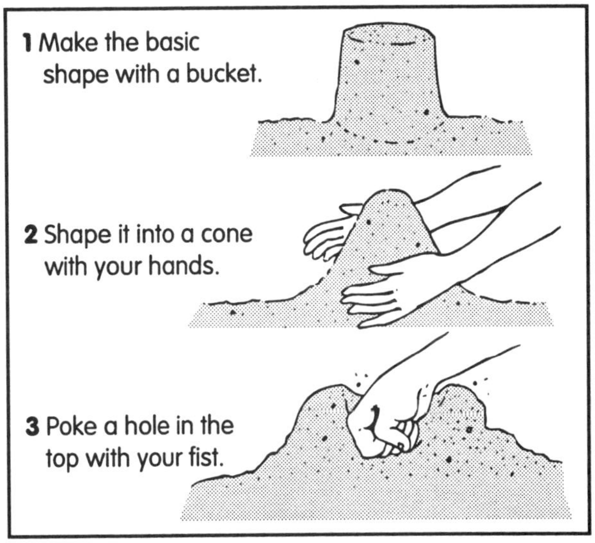

How Did the Valleys Get There?

Valleys are formed in three ways in nature.
Your students can form them the same three ways in the sandbox.

1. Some valleys are just the spaces between the mountains.

When folded mountains bend up in parallel rows, the areas between the rows become long valleys just because they didn't bend up, too.

Block mountains often occur in parallel rows too, particularly along coasts, so they have valleys between the ranges. When you see block mountain ranges that are not near the ocean, they probably used to be.

Make two folded mountain ranges. Push sand under the mountains. Recover the carpet with sand leaving the trough down the middle.

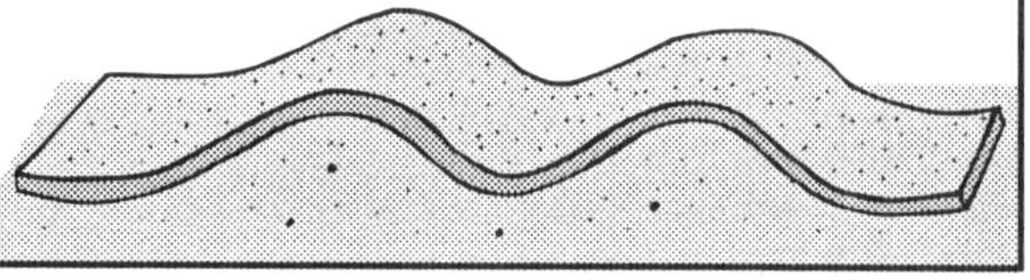

2. Some valleys are cut by rivers.

When water runs along the surface of the ground, it cuts a path and forms a river valley. River valleys are typically V-shaped with steep sides.

The biggest example of a river valley is the Grand Canyon of Arizona. It has the steep sides that tell you it was cut by a river.

Make river valleys in the sand by putting a thick layer of sand on a board. Then tilt the board and pour water down from the top with a hose or watering can.

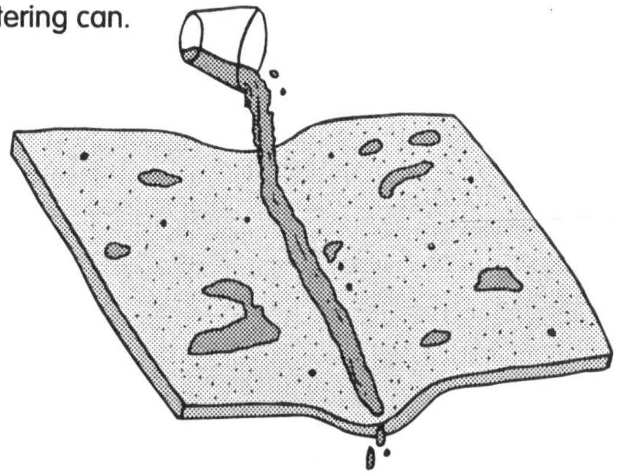

3. Some valleys are cut by glaciers (ice).

When snow falls on the ground, it usually melts in the spring or summer. In some places that are cold all year, the snow doesn't get warm enough to melt, and the snow becomes permanent ice called a glacier.

Glaciers are heavy and they slowly slide downhill, grinding big rocks beneath them into smaller rocks. These small rocks carve out valleys that tend to be U-shaped, rather than V-shaped like valleys cut by water. A well-known example of a U-shaped, glacier-formed valley is Yosemite Valley in California.

Make a glacial valley using ice cubes to carve out valleys in the sand. The cubes produce valleys of the right shape.

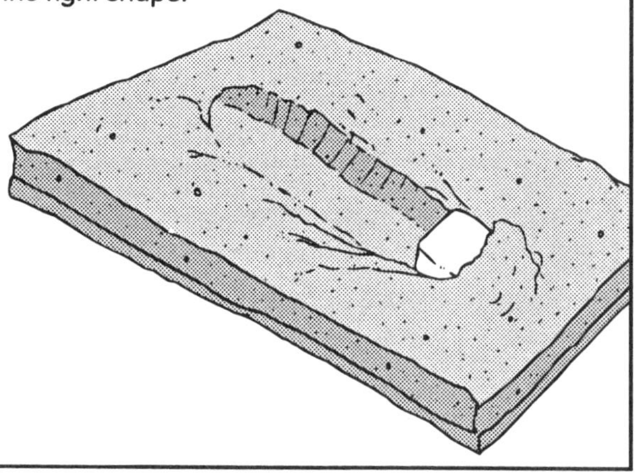

Weathering and Erosion

Weathering means breaking rock down into smaller pieces.

Jagged mountains are slowly smoothed and rounded by wind and water. It's not really the air or water that breaks the big rock into little pieces...it's the pieces of other rocks carried by the wind or water that bangs into the mountains and breaks them down.

Make a plateau (flat-topped mountain) with a bucket of sand. You can see how the straight sides go down to the flat surface below. It doesn't look very much like the real thing.

Now scrape sand from the sides and see how the shape of your plateau looks more realistic as sand collects around the base.

You can tell whether a real mountain has been worn down a lot by looking to see how much rock and sand is laying around at is base. You can see fans of soil along the edges of mountain ranges, showing where rivers have deposited the material they weathered from the mountains and then carried down to the valley.

Erosion means carrying broken pieces of rock away to a new area.

After mountains get broken into smaller and smaller pieces, those pieces can be carried off by the wind or by rivers or glaciers and end up somewhere else. This transporting of rock particles is called "erosion."

You can demonstrate water erosion in the sand-box with your hose or waterin can.

A gas-powered blower can be used to show what wind erosion is like. Be sure no one is standing where sand will blow in their eyes. Dry sand can even be blown by mouth to change the shape of the dunes. Make sure to close your eyes when making this kind of wind erosion. In fact, it's a good idea to hold your hands over your closed eyes so you won't get sand on your eyelashes, just waiting for you to open your eyes and fall, or get rubbed in.

How Are Deserts Formed?

A student can demonstrate this process for the class using a sprinkling can to represent a rain cloud moving across the land. The student can hold a cardboard cloud in front of the can as it sprinkles its way along.

Wind blows clouds from the ocean to the land carrying water that has evaporated from the ocean. Because it has so much water, you can see this cloud sprinkling lightly along the way.

When the cloud reaches the mountain range, it has to go up to get over it. The higher it goes, the colder the air is. Clouds can't hold much water in cold air, so they drop most of their remaining water on the mountains. In the winter, the water makes it all the way down to the ground as snow. In the summer, it melts on the way down and hits the ground as raindrops.

When the cloud comes down behind the mountain range, it doesn't have much water left, so the desert only gets a little bit of rain. That's why the desert is so dry. That's also how you know where to look for deserts. They develop behind mountain ranges. The important thing about deserts is not that they are hot, it's that they are *dry*.

Once in awhile a storm will have clouds that are so full of water that they make it over the mountains with lots of water left over for the desert. Because the desert doesn't have a lot of plants and good soil to hold the water, these rare desert rains result in "flash floods," and all the water runs away.

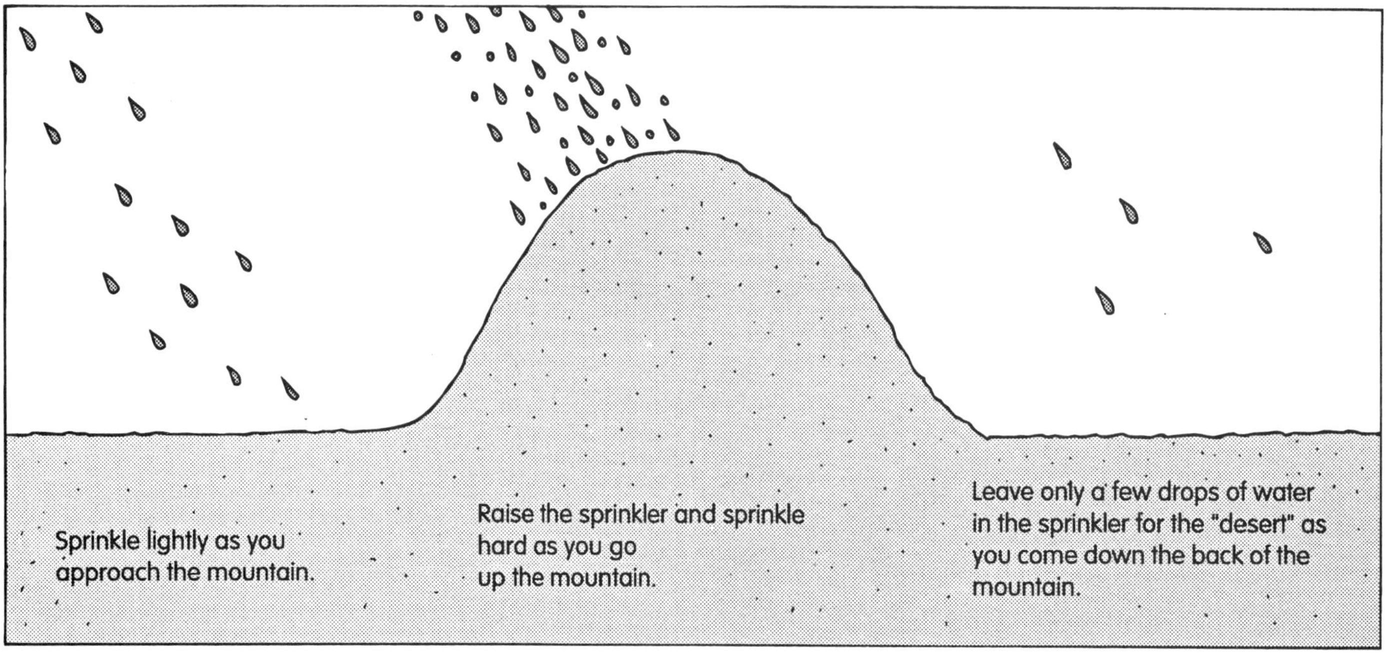

Why Do Some Rivers Wiggle?

A Cooperative Activity

Have you looked at rivers from an airplane, a hill, or a mountain? Why do some of them go straight while others look like snakes? You can experiment in the sandbox to find out for yourself why some rivers meander and others don't. Experiment with the two "variables" described here. Deep, wet sand will work best here.

How does *steepness* affect the water's course down the hill?

Spread wet sand on the surface of a board. Using a standard amount of water flowing from a hose or watering can, vary the amount of tilt of the surface, and see what different angles do to the path the water takes while flowing downhill.

Respread the sand between each trial.

Try putting gravel or rocks in the path of the water, and see what happens at different angles of tilt.

Record here what difference steepness makes. Also tell about what differences obstacles make.

____

____

____

____

____

____

____

How does the *amount of water* affect the water's course down the hill?

This time, select a medium amount of tilt and don't change it. Vary only the amount of water you pour at the top of the slope. As before, respread the sand on the surface of the board between each trial.

Again, try putting gravel or rocks in the path of the water, and see what happens with different amounts of water pouring down the slope.

Record here what difference the amount of water makes to the shape of the river. What about the obstacles with different amounts of water?

____

____

____

____

____

____

____

Making Waterways

river melted snow and rain water flowing downhill toward the sea, big enough for a boat.
pool a place in the river where the water is barely flowing and appears to be "still" water
lake where water stands but is too deep for plants to grow up through the water except at the edge
swamp a place where water collects and stands, but is so shallow that plants grow up through it
stream a small body of flowing water that often joins others to make a river.
brook a stream so shallow that the water "babbles" as it flows over the barely covered rocks
rapids a place where the river tumbles over boulders and big rocks, making "white water"
waterfall a sudden, steep place in a river where the water drops straight down to a lower level

Make a sandbox full of waterways.

1. Describe the different kinds of waterways. Make a tagboard sign for each type, cover it with contact paper and tape it to a straw. Then assign small groups to make the different kinds of waterways around the sandbox and to place the appropriate signs by their creations.

2. Have each group demonstrate their waterway, one group at a time, using the hose or watering can when it is their turn. When each group gives its demonstration, ask them to explain why they are calling theirs a brook rather than a stream, or a stream rather than a river, etc.

Show your students how to make a lake or pond using a container:

1 Dig out a hole for the container.

2 Put the container into the hole.

3 Pour water into the container.

4 Sprinkle sand into and around the container.

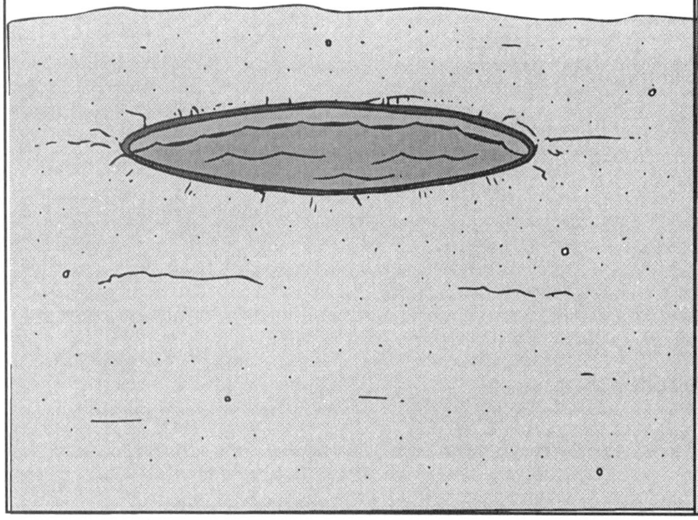

Show your students how to make a river that holds water longer:

1 Dig a small trench.

2 Line the trench with a piece of plastic trash bag.

3 Sprinkle sand into and around the trench. You could use rocks and pebbles too.

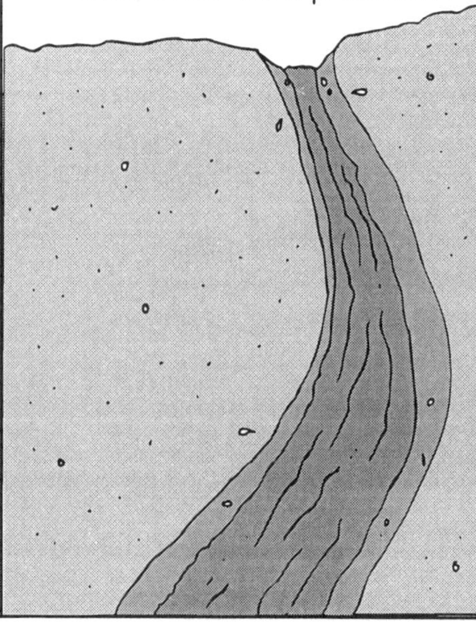

If you use plastic sheeting, be sure to keep it under your control. Just as it stops water, it stops air too, and can smother someone playing under it. Use this opportunity to talk to your students about the dangers of plastic bags for people and other living things, and that plastic bags or sheets shouldn't be played with or left lying around.

Card Games

There are two uses for the reproducible cards on page 11.

1. Cut the cards apart and use them to fill in the blank labels on the colored poster.
2. Cut apart and scramble the cards and arrange them into matching sets of three.

Note: Use these cards along with poster activities or as an independent matching activity.

river	melted snow and rain water flowing downhill toward the sea; big enough for a boat	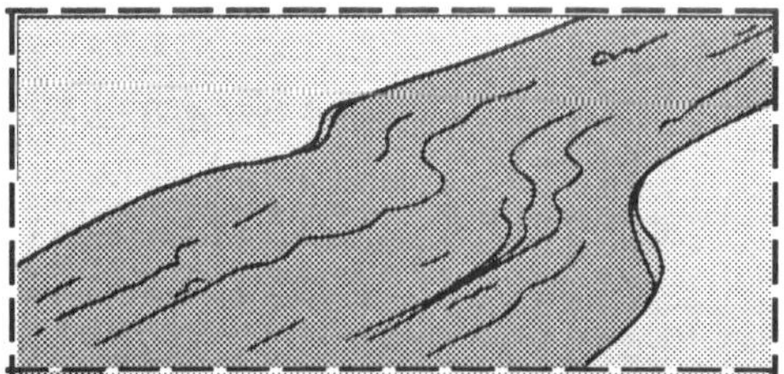
pool	a place in the river where the water is barely flowing and appears to be "still" water	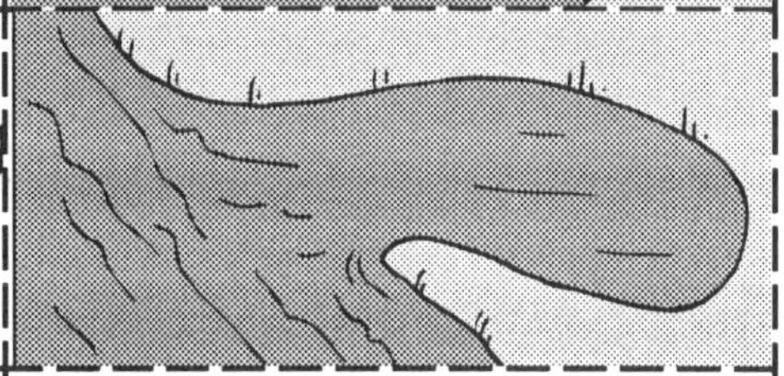
lake	where water stands but is too deep for plants to grow up through the water except at the edge	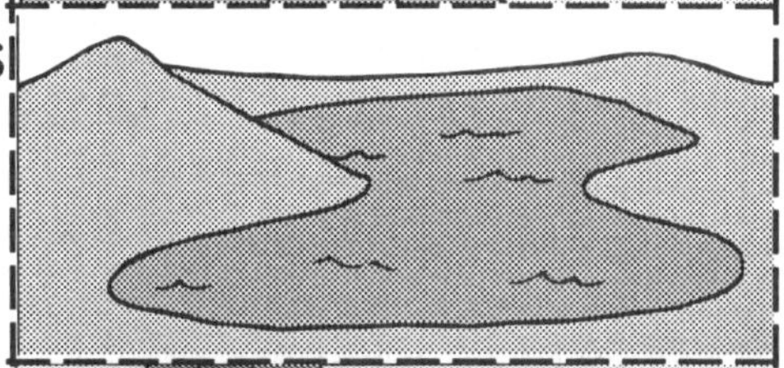
swamp	a place where water collects and stands, but is so shallow that plants grow up through it	
stream	a small body of flowing water that often joins others to make a river	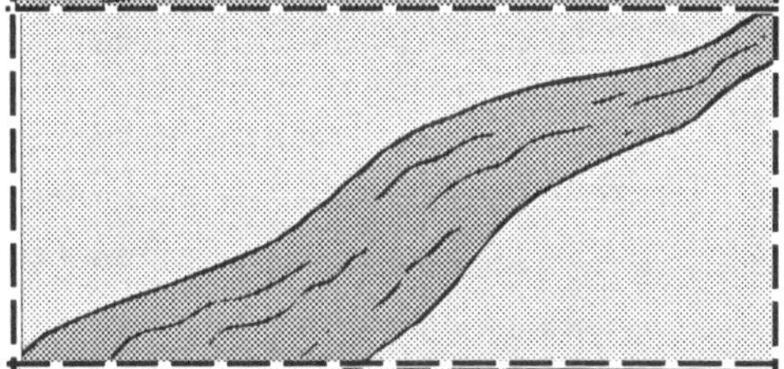
brook	a stream so shallow that the water "babbles" as it flows over the barely covered rocks	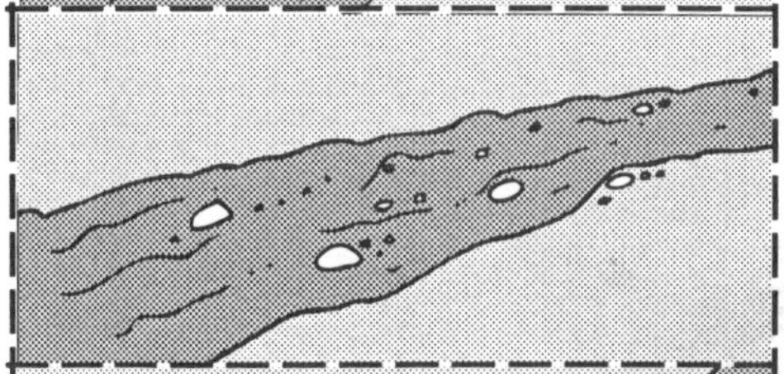
rapids	a place where a river tumbles over boulders and big rocks, making "white water"	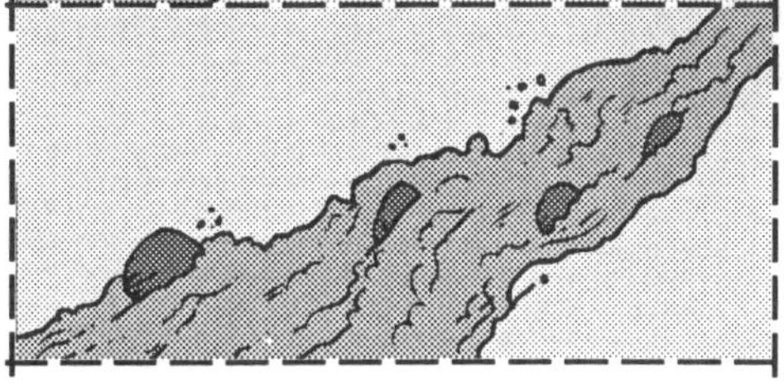
waterfall	a sudden, steep place in a river where the water falls straight down to a lower leve	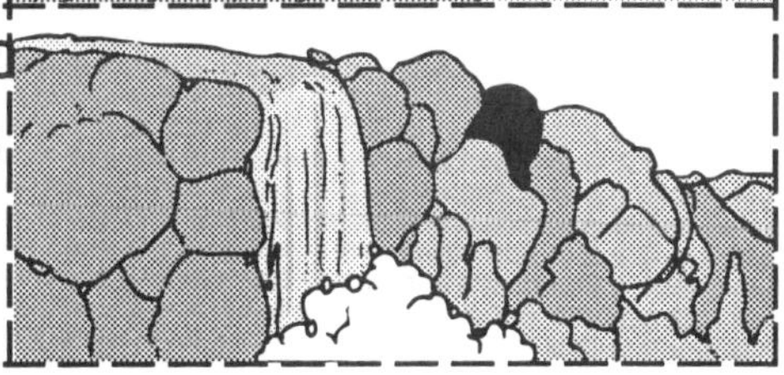

For the teacher

Making Coastal Landforms

Where does all that water wind up?

river	the collection of melted snow and rain water flowing to the sea; "fresh" (not salt) water
estuary	the place where a river meets the ocean and the water is partly "fresh", partly salty
delta	the area around a river mouth where mud from the river has been deposited and become land
ocean	a large body of water between and surrounding continents with salt water
bay	an area where salt water extends in beyond the coastline's general shape
peninsula	an area where land extends out from the coastline and is surrounded by water on three sides
island	the top of a seamount sticking above the surface, making land surrounded by water
archipelago	a string of islands that may have been a peninsula at one time, but part is under water now

Let the sandbox show the bigger picture.

1. You and your sandbox geologists are going for the big time now. Not just little rivers. Now you're going to make oceans, bays, islands, and peninsulas. Pick a group to make each of these landforms so that they become part of one big "relief map" of the area between the mountains and the sea and beyond! Prepare tagboard cards with the names of the landforms, cover them with contact paper and tape them to straws. Students will place these labels in the sandbox as they create the different forms.

2. Explain the difference between the water in:
   - a river or lake -fresh water
   - an ocean or sea - salt water
   - an estuary - partly salty

   Check to see if they understand your explanation.

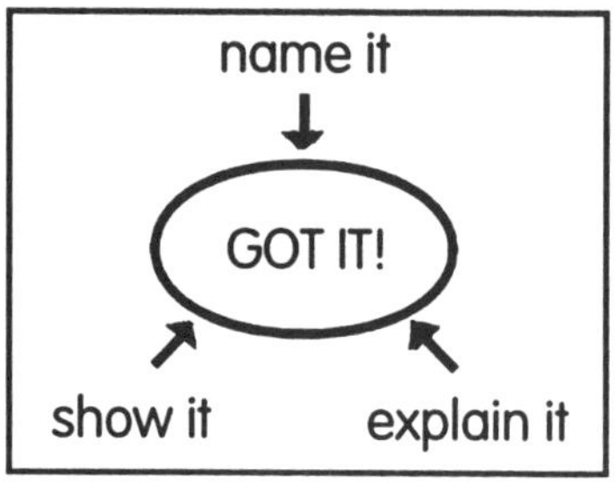

3. Make sure that your learners do more than just shape the sand. You want them to be able to name each landform and describe or define its essential characteristics. Have them demonstrate the three components of true understanding a concept:

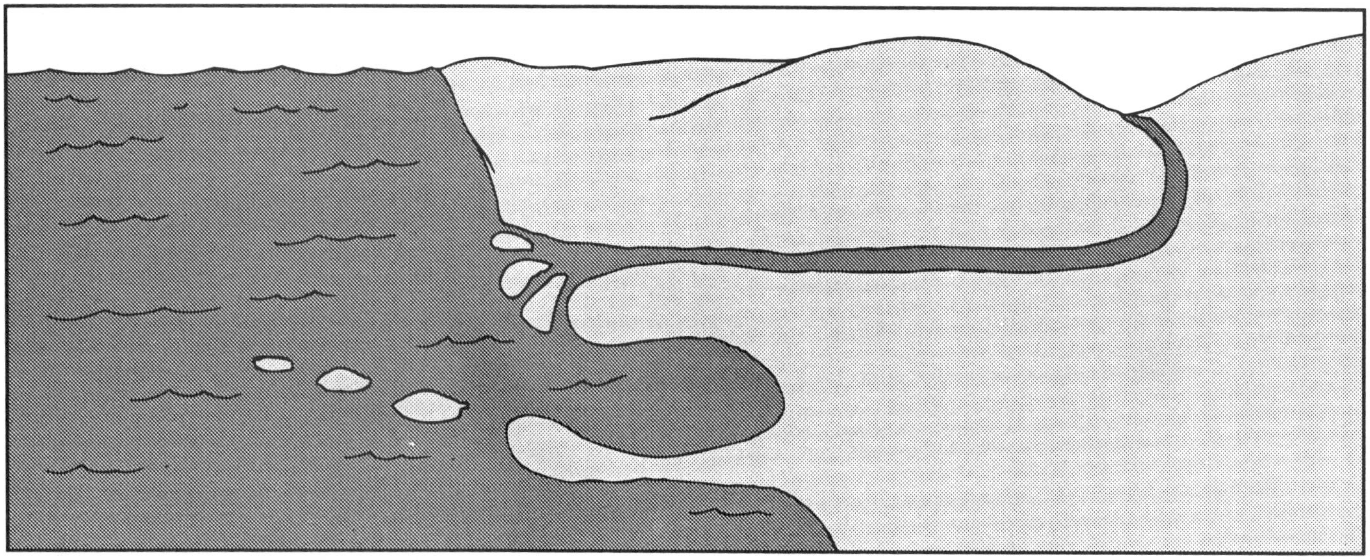

How to keep the water in your ocean...

To fill a sandbox ocean, bay, lake, etc., you can bury a plastic container or plastic sheet just under the surface of the sand. If you use plastic sheeting as described earlier for making leak proof rivers in the sandbox, make sure to explain the safety precautions on page 10. Even small pieces of plastic sheeting, such as those used for plastic bags, can be dangerous to people and animals.

Note: Use these cards along with poster activities or as an independent matching activity.

river	the collection of melted snow and rain water flowing to the sea, fresh water	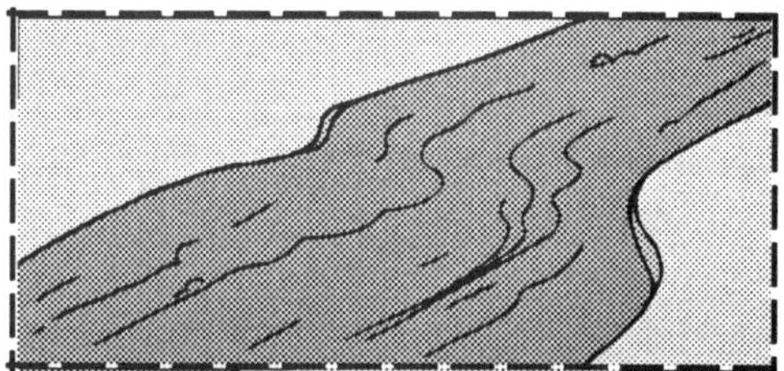
estuary	the place where a river meets the ocean and the water is partly fresh, partly salty	
delta	the area around a river mouth where mud from the river has deposited and become land	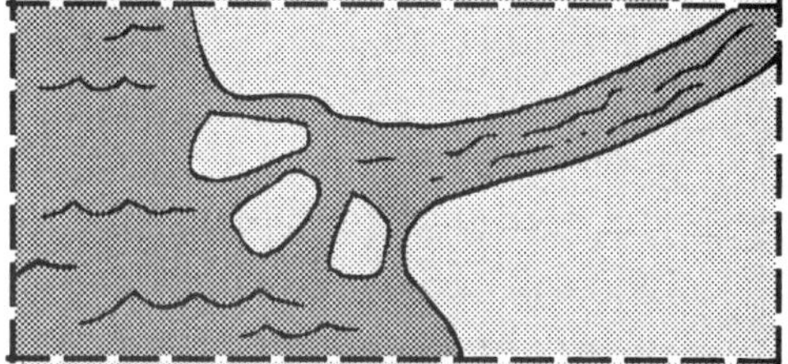
ocean	a large body of water between and surrounding continents with salt water	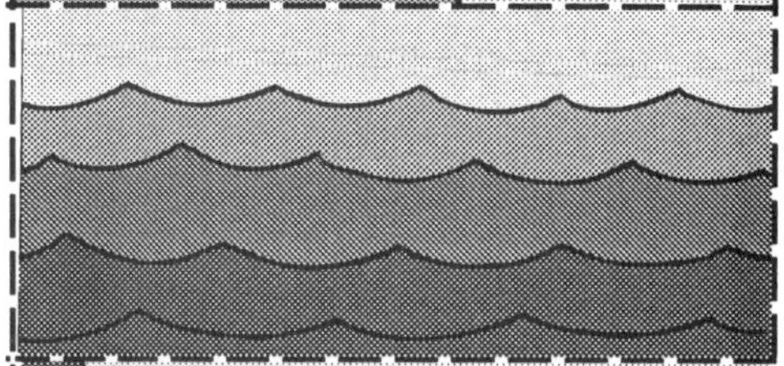
bay	an area where salt water extends in beyond the coastline's general shape	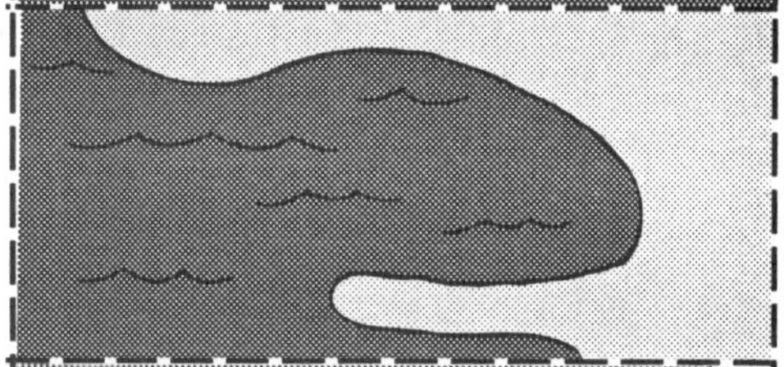
peninsula	an area where land extends out from the coastline and is surrounded by water on 3 sides	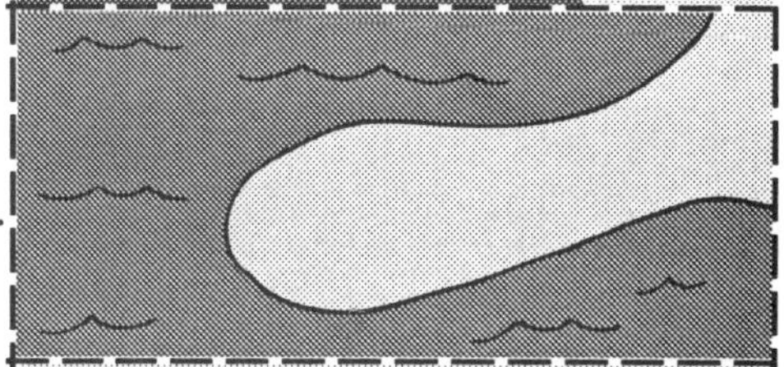
island	the top of a seamount sticking above the surface, making land surrounded by water	
archipelago	a string of islands that may have been a peninsula at one time, but part is under water now	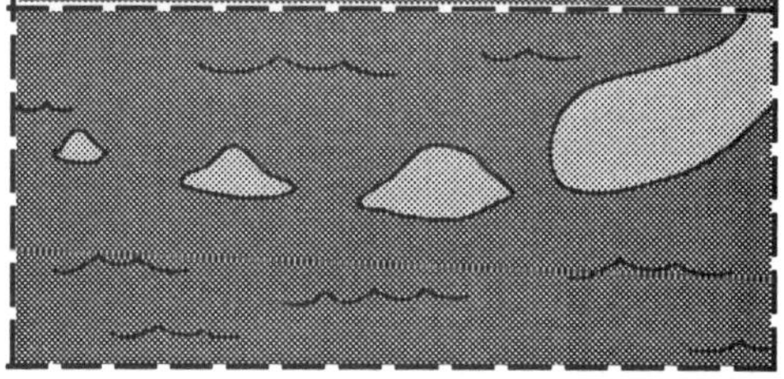

For the teacher

What do people put on land?

dam a cement wall that holds river water temporarily for later use and for making electricity
reservoir an artificial lake to save water in areas where there aren't enough natural lakes
canal an artificial river with cement sides to take water where there is no river to take it
farmland a place where the land is carved into rows of dirt and water for growing crops
town a place where the ground is covered with buildings, parking lots, and streets
highway a major street with more than one lane in each direction, often with a divider in the middle
bridge a structure to let cars and trucks or trains cross over a river, valley, or other obstruction
tunnel a tube-like passageway through a mountain or under an obstruction such as a river

A Relief Map

Turn your sandbox into a map of an imaginary place that has all of the features shown on the cutout signs on page 15. Use the signs to label your map.

This sand sculpture emphasizes the things that people do to shape the land when the natural shape doesn't have everything they want.

Discuss the pros and cons of some of the things people do to the land. Sometimes a dam is good and sometimes it isn't. What makes the difference?
How do people have dams and still provide a way for fish to make it up the river to spawn? Is there always a way to accommodate both people and nature? Whose job is it to look out for that balance?

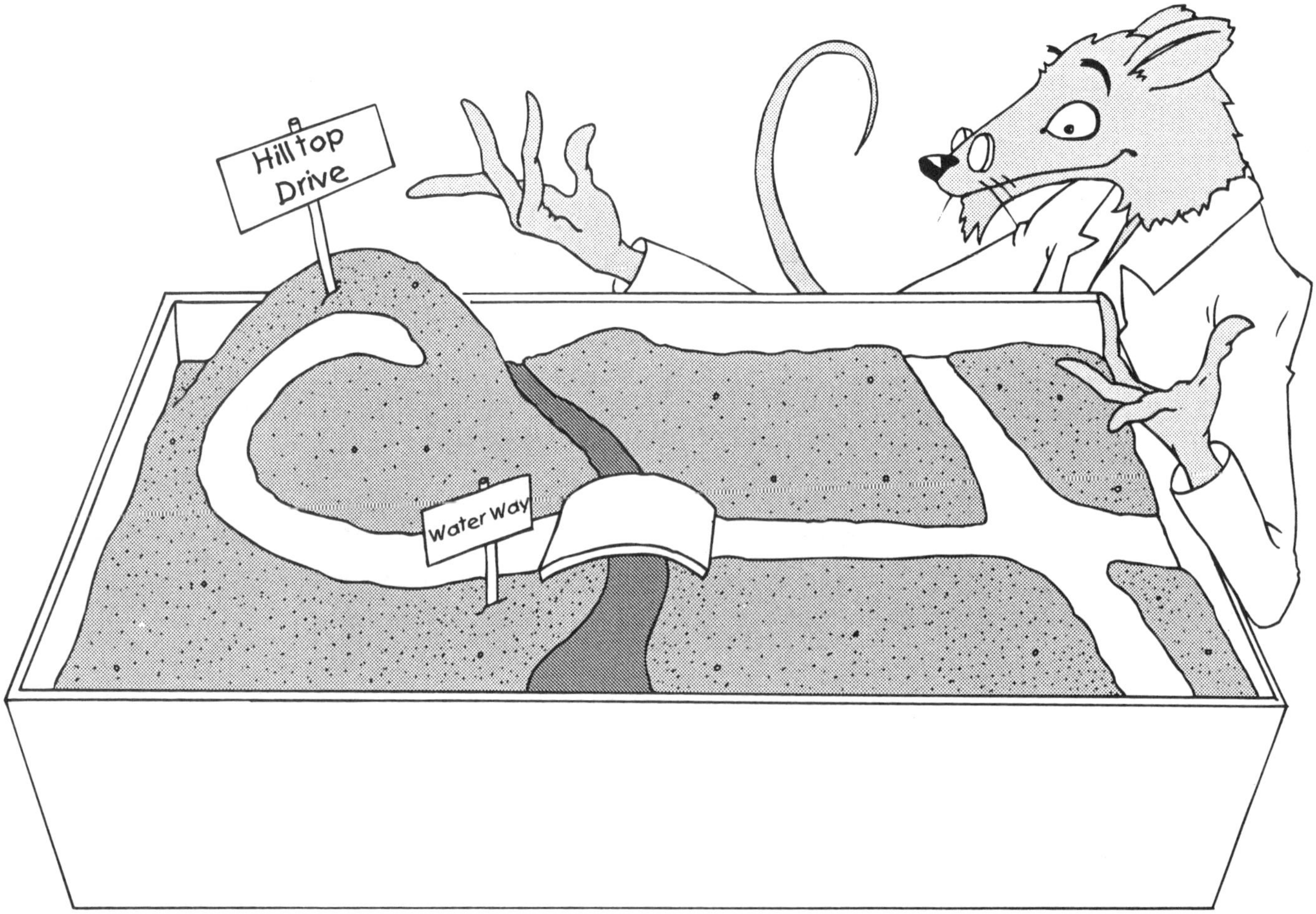

Note: Use these cards along with poster activities or as an independent matching activity.

dam	a thick wall that holds river water temporarily for later use and for making electricity	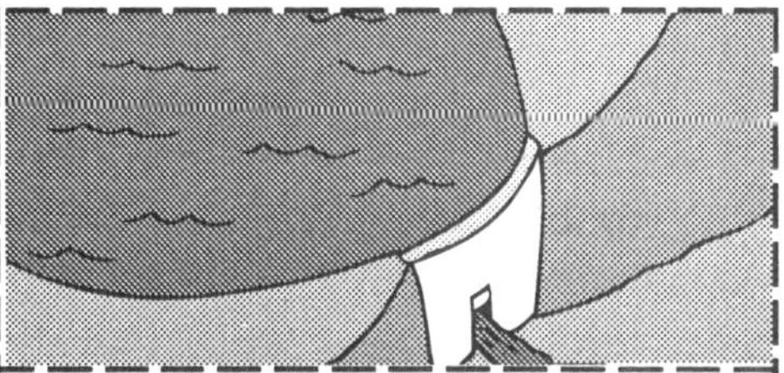
reservoir	an artificial lake to save water in areas where there aren't enough natural lakes	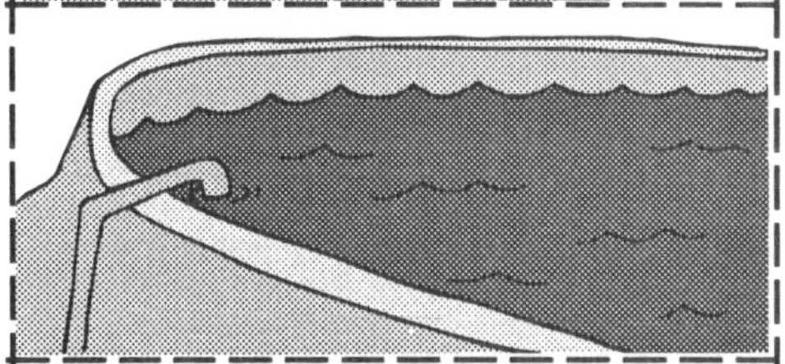
canal	an artificial river with cement sides to take water where there is no river to take it	
farmland	a place where the land is carved into rows of dirt and water for growing crops	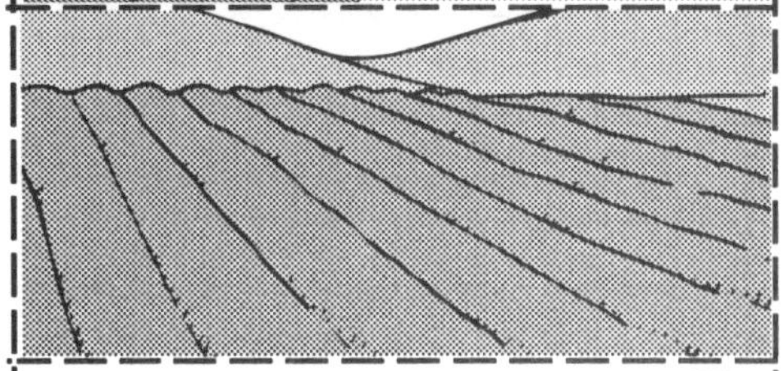
town	a place where the ground is covered with buildings, parking lots, and streets	
highway	a major street with more than one lane in each direction, often with a divider in the middle	
bridge	a structure to let cars and trucks or trains cross over a river, valley, or other obstruction	
tunnel	a tube-like passageway through a mountain or under an obstruction such as a river	

A Relief Map of Your Country or Continent

Let's take a look from way up there!

If you could get a ride in a spacecraft, you could see your whole state or province, or even your whole country. In fact, you could see other countries and whole continents. When you see a part of the world from that far away, you can only see the major features, like major mountain ranges, big rivers, giant lakes, etc.

Many schools have big maps painted on their playgrounds these days. Maybe your school has one. Those maps are usually "political" maps. They emphasize boundaries between states or countries. How about making some "physical" maps in your sandbox that emphasize the major shapes and contours of the land surface?

Make a sandbox map that shows your whole state or country. Make name signs for the major features. Try making relief maps of other countries or of continents such as Europe or Asia. Use maps and atlases for reference to show you the shapes and features. The bigger the area, the less detail you would be able to show and label. Remember...for these "relief" or "physical" maps, you won't be labelling state or country names. You'll be labeling mountain ranges, major rivers and lakes, deserts, etc.

A Physical Map

A Political Map

Can you think of ways to make your physical map realistic without messing up your sandbox? How about snow for tall mountain ranges? Could you use chopped ice? It would melt without hurting the sand. Can you use some kind of plant material to represent forested areas? It would have to be something you could take out when you were through. How about blue cloth or paper with ripples for oceans? If you do a really fancy job, you might even get your pictures in the local newspaper. Invite someone from the paper over to take a look at what you've done. Make sure to tell them what you've learned while making your maps.